I0493285

MATHadazzles

Mind Stretch Puzzles Volume 4

Reasoning with Fractions

Authors

Carole Greenes

Mary Cavanagh

Contributors: Grades 5-9 Students
Thomas Allen, Porter Aller, Mounia Bazzi, Maria deGrasse,
Steven Greenes, Samia Koudssi, Joselyn Ly, Aidan Macias,
Nickoli Mckenzie, Sebastian Moore, Ethan Organista,
Jacob Organista, Sequoia Ries, Aidan Villa

Editors
Tanner Wolfram, Senior Editor
James Kim
Jason Luc
Yifan Tian
Ping Chuan (Larry) Yong

Project Assistants
Porter Aller
Aidan Macias

Cover Design
Ping Chuan (Larry) Yong

MATHadazzles *Mind Stretch Puzzles*
Reasoning with Fractions Volume IV

MATHadazzles are number puzzles that will develop your logical reasoning abilities, your sense of numbers (different types of numbers, their characteristics, and operations with them), and your persistence in solving problems. Once you start, you won't be able to stop UNTIL you successfully solve all of the puzzles!

What is a Fraction MATHadazzle? A Fraction MATHadazzle is a 3-by-3 grid with circles at the end of each row and column. Some grid cells have clues about the numbers that will fill those cells. The numbers in the circles at the ends of the rows and the bottom of the columns are the row and column sums.

What's Your Job? Based on clues provided in some of the grid cells, you place the numbers $\frac{1}{10}, \frac{2}{10}, \frac{3}{10}, \frac{4}{10}, \frac{5}{10}, \frac{6}{10}, \frac{7}{10}, \frac{8}{10}$, and $\frac{9}{10}$ in the nine cells, so that the row and column numbers add up!

What are The Clues? Clues describe types of numbers, characteristics of those numbers (e.g. the nature of the numerator or denominator), or results of operations with the numbers. There are 12 types of clues. These may appear singly or in combination.

  Clues:

  Absolute Value: Magnitude of a number regardless of sign and recorded between two vertical line segments. Examples: $|-3| = 3$ and $|+3| = 3$.

  Decimal Numbers: Numbers written using a decimal point to show parts of a whole. Fractions can be written as decimal numbers. Examples: $\frac{1}{10} = 0.1$, $\frac{9}{10} = 0.9$, and $\frac{13}{10} = 1.3$.

  Even: Number that is divisible by 2 with a zero remainder.

  Exponent: A small number to the right and above a base number that indicates the number of times that base number should be used as a factor in multiplication. Example: $2^3 = 2 \times 2 \times 2 = 8$. In this example, 2 is the base number, 3 is the exponent, and 8 is the product.

  Factor: One of two or more numbers that produces a product. Examples: 2 and 3 are factors of 6 because $2 \times 3 = 6$, and 2 and 4 are factors of 8 because $2 \times 4 = 8$.

  Multiple: Number that is the product of a counting number and a whole number. Examples: 8 is a multiple of 4 because $2 \times 4 = 8$, and 6 is a multiple of 1 because $6 \times 1 = 6$.

  Odd: Number that is not even.

Operations: All four operations are incorporated into the clues. These are: addition, subtraction, multiplication, and division. When several different operations appear in a number sentence, then the order in which they are performed follows the Fundamental Order of Operations: Parentheses (all operations within parentheses are computed first), Powers (all base numbers with exponents), Multiplication and Division (from left to right), and finally, Addition and Subtraction (from left to right).

Prime: Number divisible by only two numbers, itself and 1. Important: Itself and 1 must be different numbers. Examples: 3 is divisible by only 1 and 3. 7 is divisible by only 1 and 7.

Square Number: Number that is the product of a non-zero number and itself. Examples: 4 is a square number because 2 x 2 = 4, and 9 is a square number because 3 x 3 = 9.

Square Root of a Number: Number that when multiplied by itself has a product equal to a given number. The square root of 16 is 4 because 4 x 4 = 16. The square root symbol is $\sqrt{}$. Example: $\sqrt{16} = 4$.

The Letter _n_: This represents any number. It is presented in phrases that give clues to the value of n. Example: $n > \frac{2}{3}$. In the set of fractions $\frac{1}{10}$ through $\frac{9}{10}$, n could be $\frac{5}{10}, \frac{6}{10}, \frac{7}{10}, \frac{8}{10}$, or $\frac{9}{10}$. Another example: $n < \frac{1}{2}$, so n could be $\frac{1}{10}, \frac{2}{10}, \frac{3}{10}$, or $\frac{4}{10}$.

If you are interested in improving your dazzling solution talents, consider getting Volumes I, II and III with more puzzles. Answers are at the back of each book.

Enjoy Solving!

Your MATHadazzling Authors, Contributors, Editors and Project Assistants

1

Put these numbers in the squares $\frac{1}{10}$ $\frac{2}{10}$ $\frac{3}{10}$ $\frac{4}{10}$ $\frac{5}{10}$ $\frac{6}{10}$ $\frac{7}{10}$ $\frac{8}{10}$ $\frac{9}{10}$

Add across $\longrightarrow$

Add down $\downarrow$

Sums are in $\bigcirc$

$1 - \frac{3}{5}$			$\left(1\frac{2}{5}\right)$
		$\frac{2}{10} \times 4$	$\left(1\frac{4}{5}\right)$
$6 \div 10 \div 3$	$\frac{2}{10} + \frac{3}{10}$		$\left(1\frac{3}{10}\right)$

$\left(1\frac{3}{10}\right)$ $\left(1\frac{7}{10}\right)$ $\left(1\frac{1}{2}\right)$

Put these numbers in the squares $\frac{1}{10}$ $\frac{2}{10}$ $\frac{3}{10}$ $\frac{4}{10}$ $\frac{5}{10}$ $\frac{6}{10}$ $\frac{7}{10}$ $\frac{8}{10}$ $\frac{9}{10}$

Add across $\longrightarrow$

Add down $\downarrow$

Sums are in $\bigcirc$

$\frac{3}{10} \times 3$			$\left(1\frac{3}{5}\right)$
	$\frac{2}{10} + \frac{3}{10}$		$\left(1\frac{3}{5}\right)$
$1 - \frac{4}{5}$		$1 - \frac{3}{10}$	$\left(1\frac{3}{10}\right)$
$\left(1\frac{2}{5}\right)$	$\left(1\frac{1}{2}\right)$	$\left(1\frac{3}{5}\right)$	

Put these numbers in the squares $\frac{1}{10}$ $\frac{2}{10}$ $\frac{3}{10}$ $\frac{4}{10}$ $\frac{5}{10}$ $\frac{6}{10}$ $\frac{7}{10}$ $\frac{8}{10}$ $\frac{9}{10}$

Add across $\longrightarrow$

Add down $\downarrow$

Sums are in $\bigcirc$

$\frac{3}{5} \times \frac{3}{2}$		0.7
0.5	$1 - \frac{9}{10}$	

$2\frac{1}{5}$

$\frac{4}{5}$

$1\frac{1}{2}$

$1\frac{4}{5}$ 1 $1\frac{7}{10}$

Put these numbers in the squares $\frac{1}{10}$ $\frac{2}{10}$ $\frac{3}{10}$ $\frac{4}{10}$ $\frac{5}{10}$ $\frac{6}{10}$ $\frac{7}{10}$ $\frac{8}{10}$ $\frac{9}{10}$

Add across $\longrightarrow$

Add down $\downarrow$

Sums are in $\bigcirc$

	$4 \div 5$	$\sqrt{81} \div 10$	$2\frac{1}{10}$
$\frac{4}{5} \div \frac{16}{20} \times \frac{1}{10}$			$1\frac{2}{5}$
		$1 \div 2$	1
$\frac{7}{10}$	$1\frac{7}{10}$	$2\frac{1}{10}$	

5

Put these numbers in the squares $\frac{1}{10}$ $\frac{2}{10}$ $\frac{3}{10}$ $\frac{4}{10}$ $\frac{5}{10}$ $\frac{6}{10}$ $\frac{7}{10}$ $\frac{8}{10}$ $\frac{9}{10}$

Add across →

Add down ↓

Sums are in ◯

	$18 \div 3 \div 10$	$3^3 \div 10$	$2\frac{1}{5}$
	$2 \div 10$	$\sqrt{64} \div 10$	$1\frac{1}{10}$
			$1\frac{1}{5}$
$1\frac{3}{10}$	$1\frac{1}{5}$	2	

6

Put these numbers in the squares $\frac{1}{10}$ $\frac{2}{10}$ $\frac{3}{10}$ $\frac{4}{10}$ $\frac{5}{10}$ $\frac{6}{10}$ $\frac{7}{10}$ $\frac{8}{10}$ $\frac{9}{10}$

Add across $\longrightarrow$

Add down $\downarrow$

Sums are in $\bigcirc$

$96 \div 24 \div 10$	$3^2 \div 10$		$1\frac{9}{10}$
$\sqrt{64} \div 10$			$1\frac{2}{5}$
		$\sqrt{4} \div \sqrt{100}$	$1\frac{1}{5}$
$1\frac{1}{2}$	$1\frac{7}{10}$	$1\frac{3}{10}$	

Put these numbers in the squares $\frac{1}{10}$ $\frac{2}{10}$ $\frac{3}{10}$ $\frac{4}{10}$ $\frac{5}{10}$ $\frac{6}{10}$ $\frac{7}{10}$ $\frac{8}{10}$ $\frac{9}{10}$

Add across $\longrightarrow$

Add down $\downarrow$

Sums are in $\bigcirc$

$\dfrac{(8-2) \div 1.5}{10}$			$1\frac{1}{5}$
$\sqrt{1} \div 10$	$1 \div 2$		$\frac{9}{10}$
		$\dfrac{7}{10} < n < \dfrac{9}{10}$	$2\frac{2}{5}$
$1\frac{2}{5}$	$1\frac{4}{5}$	$1\frac{3}{10}$	

8

Put these numbers in the squares $\frac{1}{10}$ $\frac{2}{10}$ $\frac{3}{10}$ $\frac{4}{10}$ $\frac{5}{10}$ $\frac{6}{10}$ $\frac{7}{10}$ $\frac{8}{10}$ $\frac{9}{10}$

Add across $\longrightarrow$

Add down $\downarrow$

Sums are in $\bigcirc$

		$3 \div 5$	$1\frac{1}{2}$
$\dfrac{800 \div 200}{10}$		$\sqrt{64} \div 10$	$1\frac{7}{10}$
$\dfrac{\sqrt{25} - \sqrt{16}}{10}$			$1\frac{3}{10}$
$1\frac{1}{5}$	1	$2\frac{3}{10}$	

Put these numbers in the squares $\frac{1}{10}$ $\frac{2}{10}$ $\frac{3}{10}$ $\frac{4}{10}$ $\frac{5}{10}$ $\frac{6}{10}$ $\frac{7}{10}$ $\frac{8}{10}$ $\frac{9}{10}$

Add across $\longrightarrow$

Add down $\downarrow$

Sums are in $\bigcirc$

	$72 \div 24 \div 10$		$1\frac{4}{5}$
		$\sqrt{100} \times \frac{1}{100}$	$1\frac{3}{5}$
$\sqrt{64} \div \sqrt{400}$	$\dfrac{\text{Even Prime}}{10}$		$1\frac{1}{10}$
$1\frac{4}{5}$	$1\frac{2}{5}$	$1\frac{3}{10}$	

10

Put these numbers in the squares $\frac{1}{10}$ $\frac{2}{10}$ $\frac{3}{10}$ $\frac{4}{10}$ $\frac{5}{10}$ $\frac{6}{10}$ $\frac{7}{10}$ $\frac{8}{10}$ $\frac{9}{10}$

Add across $\longrightarrow$

Add down $\downarrow$

Sums are in $\bigcirc$

	$\frac{1+1+2}{10}$	$\frac{1.5 \times 2 + 3}{10}$	$1\frac{9}{10}$
			1
$\frac{1}{2} + \frac{1}{5}$		$\frac{6 \times 0 \times 5 + 1}{10}$	$1\frac{3}{5}$
$1\frac{4}{5}$	$1\frac{1}{2}$	$1\frac{1}{5}$	

Put these numbers in the squares $\frac{1}{10}$ $\frac{2}{10}$ $\frac{3}{10}$ $\frac{4}{10}$ $\frac{5}{10}$ $\frac{6}{10}$ $\frac{7}{10}$ $\frac{8}{10}$ $\frac{9}{10}$

Add across $\longrightarrow$

Add down $\downarrow$

Sums are in $\bigcirc$

			$1\frac{1}{10}$
$4 \div 5$		$\frac{10}{100} + \frac{300}{1,000}$	$2\frac{1}{10}$
	$0.7 - \frac{6}{10}$		$1\frac{3}{10}$
$1\frac{3}{5}$	$1\frac{3}{5}$	$1\frac{3}{10}$	

12

Put these numbers in the squares $\frac{1}{10}$ $\frac{2}{10}$ $\frac{3}{10}$ $\frac{4}{10}$ $\frac{5}{10}$ $\frac{6}{10}$ $\frac{7}{10}$ $\frac{8}{10}$ $\frac{9}{10}$

Add across →

Add down ↓

Sums are in ◯

$\frac{\text{Multiple of } 9}{10}$		$6 \div 2 \div 10$
$2^2 \div 10$		$10 \div 2 \div 10$

Right circles (top to bottom): $1\frac{3}{10}$, $1\frac{1}{2}$, $1\frac{7}{10}$

Bottom circles (left to right): $1\frac{1}{2}$, $1\frac{1}{2}$, $1\frac{1}{2}$

13

Put these numbers in the squares $\frac{1}{10}$ $\frac{2}{10}$ $\frac{3}{10}$ $\frac{4}{10}$ $\frac{5}{10}$ $\frac{6}{10}$ $\frac{7}{10}$ $\frac{8}{10}$ $\frac{9}{10}$

Add across $\longrightarrow$

Add down $\downarrow$

Sums are in $\bigcirc$

$\frac{1}{5} + \frac{1}{5}$			$1\frac{4}{5}$
$\frac{8}{5} - \frac{4}{5}$			$1\frac{1}{2}$
		$\sqrt{4} \div 10$	$1\frac{1}{5}$
$1\frac{1}{2}$	$1\frac{7}{10}$	$1\frac{3}{10}$	

Put these numbers in the squares $\frac{1}{10}$ $\frac{2}{10}$ $\frac{3}{10}$ $\frac{4}{10}$ $\frac{5}{10}$ $\frac{6}{10}$ $\frac{7}{10}$ $\frac{8}{10}$ $\frac{9}{10}$

Add across $\longrightarrow$

Add down $\downarrow$

Sums are in $\bigcirc$

	$\dfrac{\text{Odd Prime}}{10}$	$\dfrac{\text{Triangular}}{10}$	
			$1\frac{1}{2}$
	$1 \div 10$	$4 \times 2 \div 10$	$1\frac{3}{10}$
$\dfrac{3 \times 6 - 9}{10}$			$1\frac{7}{10}$

$\bigcirc 2$ $\qquad$ $\bigcirc \dfrac{4}{5}$ $\qquad$ $\bigcirc 1\frac{7}{10}$

15

Put these numbers in the squares $\frac{1}{10}$ $\frac{2}{10}$ $\frac{3}{10}$ $\frac{4}{10}$ $\frac{5}{10}$ $\frac{6}{10}$ $\frac{7}{10}$ $\frac{8}{10}$ $\frac{9}{10}$

Add across ➔

Add down ↓

Sums are in ◯

$\sqrt{9} \div 10$		$\sqrt{81} \div 10$	$1\frac{2}{5}$
			$1\frac{3}{10}$
$0.1 \times 10 - 0.5$		$\dfrac{4 \times 4 - 3^2}{10}$	$1\frac{4}{5}$
$1\frac{1}{5}$	$\frac{9}{10}$	$2\frac{2}{5}$	

16

Put these numbers in the squares $\frac{1}{10}$ $\frac{2}{10}$ $\frac{3}{10}$ $\frac{4}{10}$ $\frac{5}{10}$ $\frac{6}{10}$ $\frac{7}{10}$ $\frac{8}{10}$ $\frac{9}{10}$

Add across $\longrightarrow$

Add down $\downarrow$

Sums are in $\bigcirc$

			$1\frac{2}{5}$
	$\frac{1}{2}+\frac{2}{5}$	$2^3 \div 10$	$2\frac{3}{10}$
0.3		$\dfrac{\text{Triangular}}{10}$	$\frac{4}{5}$
$1\frac{3}{5}$	$1\frac{1}{2}$	$1\frac{2}{5}$	

Put these numbers in the squares $\frac{1}{10}$ $\frac{2}{10}$ $\frac{3}{10}$ $\frac{4}{10}$ $\frac{5}{10}$ $\frac{6}{10}$ $\frac{7}{10}$ $\frac{8}{10}$ $\frac{9}{10}$

Add across $\longrightarrow$

Add down $\downarrow$

Sums are in $\bigcirc$

$8 \times \frac{1}{10}$	$\sqrt{25} \div 10$		2
	$\frac{9}{10} - \frac{4}{5}$		$\frac{3}{5}$
		$4^2 \div 4 \div 10$	$1\frac{9}{10}$
$1\frac{7}{10}$	$1\frac{1}{2}$	$1\frac{3}{10}$	

Put these numbers in the squares $\frac{1}{10}$ $\frac{2}{10}$ $\frac{3}{10}$ $\frac{4}{10}$ $\frac{5}{10}$ $\frac{6}{10}$ $\frac{7}{10}$ $\frac{8}{10}$ $\frac{9}{10}$

Add across $\longrightarrow$

Add down $\downarrow$

Sums are in $\bigcirc$

$36 \div 6 \div 10$			$1\frac{1}{2}$
	$\sqrt{4} \times \frac{1}{10}$	$25 \div 50$	1
		$3 \times 3 \div \sqrt{100}$	2
$1\frac{3}{5}$	$1\frac{2}{5}$	$1\frac{1}{2}$	

Put these numbers in the squares $\frac{1}{10}$ $\frac{2}{10}$ $\frac{3}{10}$ $\frac{4}{10}$ $\frac{5}{10}$ $\frac{6}{10}$ $\frac{7}{10}$ $\frac{8}{10}$ $\frac{9}{10}$

Add across ⟶

Add down ↓

Sums are in ◯

$4 \div 5$		$\dfrac{\text{Triangular}}{10}$
	$\dfrac{2}{5} + \dfrac{1}{2}$	
$0 \times \sqrt{100} + \dfrac{1}{10}$		

Circles (right side, top to bottom): $1\frac{9}{10}$, $1\frac{3}{5}$, 1

Circles (bottom, left to right): $1\frac{1}{5}$, $2\frac{1}{10}$, $1\frac{1}{5}$

Put these numbers in the squares $\frac{1}{10}$ $\frac{2}{10}$ $\frac{3}{10}$ $\frac{4}{10}$ $\frac{5}{10}$ $\frac{6}{10}$ $\frac{7}{10}$ $\frac{8}{10}$ $\frac{9}{10}$

Add across $\longrightarrow$

Add down $\downarrow$

Sums are in $\bigcirc$

			$1\frac{1}{5}$
$4 \div 5$		$3 \div 5$	$1\frac{9}{10}$
	$\frac{1}{2} + \frac{1}{5}$		$1\frac{2}{5}$
$1\frac{3}{10}$	$1\frac{2}{5}$	$1\frac{4}{5}$	

21

Put these numbers in the squares $\frac{1}{10}$ $\frac{2}{10}$ $\frac{3}{10}$ $\frac{4}{10}$ $\frac{5}{10}$ $\frac{6}{10}$ $\frac{7}{10}$ $\frac{8}{10}$ $\frac{9}{10}$

Add across $\longrightarrow$

Add down $\downarrow$

Sums are in $\bigcirc$

	$8 \times \frac{1}{10}$		$1\frac{1}{2}$
$\frac{1}{2} + \frac{1}{5}$			2
	$0.3 - \frac{2}{10}$	$\frac{\text{Odd}}{10}$	1
$1\frac{1}{2}$	$1\frac{3}{10}$	$1\frac{7}{10}$	

22

Put these numbers in the squares $\frac{1}{10}$ $\frac{2}{10}$ $\frac{3}{10}$ $\frac{4}{10}$ $\frac{5}{10}$ $\frac{6}{10}$ $\frac{7}{10}$ $\frac{8}{10}$ $\frac{9}{10}$

Add across →

Add down ↓

Sums are in ◯

0.1	$\frac{\text{Triangular}}{10}$		$1\frac{1}{10}$
		$32 \div 4 \div 10$	$1\frac{4}{5}$
$2.5 \times 2 \div 10$			$1\frac{3}{5}$
$1\frac{3}{10}$	$1\frac{4}{5}$	$1\frac{2}{5}$	

Put these numbers in the squares $\frac{1}{10}$ $\frac{2}{10}$ $\frac{3}{10}$ $\frac{4}{10}$ $\frac{5}{10}$ $\frac{6}{10}$ $\frac{7}{10}$ $\frac{8}{10}$ $\frac{9}{10}$

Add across →

Add down ↓

Sums are in ◯

$8 \times \frac{1}{10}$	$\frac{1}{2} + \frac{1}{5}$		$2\frac{1}{10}$
		$2 - 1.1$	$1\frac{1}{2}$
$n > \frac{2}{5}$			$\frac{9}{10}$
$1\frac{7}{10}$	$1\frac{1}{5}$	$1\frac{3}{5}$	

Put these numbers in the squares $\frac{1}{10}$ $\frac{2}{10}$ $\frac{3}{10}$ $\frac{4}{10}$ $\frac{5}{10}$ $\frac{6}{10}$ $\frac{7}{10}$ $\frac{8}{10}$ $\frac{9}{10}$

Add across $\longrightarrow$

Add down $\downarrow$

Sums are in $\bigcirc$

$\sqrt{100} \times \frac{1}{10} \div 10$		$\frac{1}{2} + \frac{1}{5}$	$1\frac{7}{10}$
			$1\frac{1}{2}$
$n < \frac{1}{2}$		$8 \times \frac{1}{10}$	$1\frac{3}{10}$
1	$1\frac{3}{5}$	$1\frac{9}{10}$	

Put these numbers in the squares $\frac{1}{10}$ $\frac{2}{10}$ $\frac{3}{10}$ $\frac{4}{10}$ $\frac{5}{10}$ $\frac{6}{10}$ $\frac{7}{10}$ $\frac{8}{10}$ $\frac{9}{10}$

Add across $\longrightarrow$

Add down $\downarrow$

Sums are in $\bigcirc$

			$1\frac{9}{10}$
$\sqrt{36} \div 10$	$\dfrac{\text{Square}}{10}$		$1\frac{1}{10}$
	$\dfrac{2}{5} + \dfrac{3}{10}$	$\dfrac{1}{3} + \dfrac{1}{6}$	$1\frac{1}{2}$
$1\frac{1}{10}$	2	$1\frac{2}{5}$	

26

Put these numbers in the squares $\frac{1}{10}$ $\frac{2}{10}$ $\frac{3}{10}$ $\frac{4}{10}$ $\frac{5}{10}$ $\frac{6}{10}$ $\frac{7}{10}$ $\frac{8}{10}$ $\frac{9}{10}$

Add across $\longrightarrow$

Add down $\downarrow$

Sums are in $\bigcirc$

$\dfrac{	-2	}{10}+\dfrac{	+2	}{10}$		$\dfrac{\text{Odd Prime}}{10}$	$1\dfrac{3}{10}$
	$\sqrt{25}\div 10$		$1\dfrac{9}{10}$
0.1			$1\dfrac{3}{10}$
$1\dfrac{1}{10}$	1	$2\dfrac{2}{5}$	

Put these numbers in the squares $\frac{1}{10}$ $\frac{2}{10}$ $\frac{3}{10}$ $\frac{4}{10}$ $\frac{5}{10}$ $\frac{6}{10}$ $\frac{7}{10}$ $\frac{8}{10}$ $\frac{9}{10}$

Add across →

Add down ↓

Sums are in ◯

	$\frac{\text{Odd}}{10}$	$3^2 \div 10$	
			$1\frac{4}{5}$
0.3	$\frac{1}{2} + \frac{1}{5}$		$1\frac{2}{5}$
			$1\frac{3}{10}$

$1\frac{7}{10}$ 1 $1\frac{4}{5}$

Put these numbers in the squares $\frac{1}{10}$ $\frac{2}{10}$ $\frac{3}{10}$ $\frac{4}{10}$ $\frac{5}{10}$ $\frac{6}{10}$ $\frac{7}{10}$ $\frac{8}{10}$ $\frac{9}{10}$

Add across →

Add down ↓

Sums are in ◯

		$1 \div 2$
$\frac{2}{5} + \frac{1}{5} + \frac{1}{5}$	$2 \div 5 \div 2$	
	$2^2 \div \sqrt{100}$	

Row sums: $\frac{9}{10}$, $1\frac{3}{5}$, 2

Column sums: $1\frac{4}{5}$, $\frac{9}{10}$, $1\frac{4}{5}$

Put these numbers in the squares $\frac{1}{10}$ $\frac{2}{10}$ $\frac{3}{10}$ $\frac{4}{10}$ $\frac{5}{10}$ $\frac{6}{10}$ $\frac{7}{10}$ $\frac{8}{10}$ $\frac{9}{10}$

Add across $\longrightarrow$

Add down $\downarrow$

Sums are in $\bigcirc$

$\sqrt{9} \div 10$		$\dfrac{\text{Multiple of 5}}{10}$
	$\dfrac{1}{2} + \dfrac{1}{5}$	
		$6 \div 10$

$1\dfrac{7}{10}$

1

$1\dfrac{4}{5}$

$1\dfrac{1}{5}$ 2 $1\dfrac{3}{10}$

Put these numbers in the squares $\frac{1}{10}$ $\frac{2}{10}$ $\frac{3}{10}$ $\frac{4}{10}$ $\frac{5}{10}$ $\frac{6}{10}$ $\frac{7}{10}$ $\frac{8}{10}$ $\frac{9}{10}$

Add across →

Add down ↓

Sums are in ◯

	$\sqrt{9} \div 10$		$1\frac{1}{2}$
$\frac{1}{5} \times 3$			$1\frac{1}{5}$
	$3.6 \div 4$	$\dfrac{Prime}{10}$	$1\frac{4}{5}$
$1\frac{1}{5}$	$1\frac{3}{10}$	2	

31

Put these numbers in the squares $\frac{1}{10}$ $\frac{2}{10}$ $\frac{3}{10}$ $\frac{4}{10}$ $\frac{5}{10}$ $\frac{6}{10}$ $\frac{7}{10}$ $\frac{8}{10}$ $\frac{9}{10}$

Add across ⟶

Add down ↓

Sums are in ◯

			$1\frac{4}{5}$
2 x 0.25	$\dfrac{\text{Triangular}}{10}$		$1\frac{1}{5}$
0.1		$0.4 + 0.3 + \dfrac{1}{10}$	$1\frac{1}{2}$
$\frac{4}{5}$	$1\frac{3}{5}$	$2\frac{1}{10}$	

32

Put these numbers in the squares $\frac{1}{10}$ $\frac{2}{10}$ $\frac{3}{10}$ $\frac{4}{10}$ $\frac{5}{10}$ $\frac{6}{10}$ $\frac{7}{10}$ $\frac{8}{10}$ $\frac{9}{10}$

Add across $\longrightarrow$

Add down $\downarrow$

Sums are in $\bigcirc$

$\frac{Prime}{10}$			$\left(1\frac{3}{5}\right)$
$\frac{3}{10}$ x 2		$8 \div 10$	$\left(1\frac{1}{2}\right)$
	$\frac{1}{2} + \frac{1}{5}$		$\left(1\frac{2}{5}\right)$
$\left(1\frac{1}{5}\right)$	$\left(1\frac{7}{10}\right)$	$\left(1\frac{3}{5}\right)$	

Put these numbers in the squares $\frac{1}{10}$ $\frac{2}{10}$ $\frac{3}{10}$ $\frac{4}{10}$ $\frac{5}{10}$ $\frac{6}{10}$ $\frac{7}{10}$ $\frac{8}{10}$ $\frac{9}{10}$

Add across $\longrightarrow$

Add down $\downarrow$

Sums are in $\bigcirc$

		0.5
	$\sqrt{81} \div 10$	
$\frac{5}{10} + \frac{1}{5}$		$\frac{\text{Multiple of 3}}{10}$

Right circles: $1\frac{2}{5}$, $1\frac{2}{5}$, $1\frac{7}{10}$

Bottom circles: $1\frac{7}{10}$, $1\frac{2}{5}$, $1\frac{2}{5}$

Put these numbers in the squares $\frac{1}{10}$ $\frac{2}{10}$ $\frac{3}{10}$ $\frac{4}{10}$ $\frac{5}{10}$ $\frac{6}{10}$ $\frac{7}{10}$ $\frac{8}{10}$ $\frac{9}{10}$

Add across ⟶

Add down ↓

Sums are in ◯

			2
$\sqrt{9} \times 9 \div 9 \div 10$		$\frac{1}{10} + \frac{1}{10}$	$\frac{3}{5}$
	$2 \times 4 \div \sqrt{100}$		$1\frac{9}{10}$
$1\frac{4}{5}$	$1\frac{3}{5}$	$1\frac{1}{10}$	

Put these numbers in the squares $\frac{1}{10}$ $\frac{2}{10}$ $\frac{3}{10}$ $\frac{4}{10}$ $\frac{5}{10}$ $\frac{6}{10}$ $\frac{7}{10}$ $\frac{8}{10}$ $\frac{9}{10}$

Add across $\longrightarrow$

Add down $\downarrow$

Sums are in $\bigcirc$

	$8 \times \frac{1}{10}$		$1\frac{7}{10}$
$5 - 4.3$			$2\frac{1}{10}$
	$4^2 \div 4 \div 10$		$\frac{7}{10}$
$1\frac{1}{5}$	$2\frac{1}{10}$	$1\frac{1}{5}$	

Put these numbers in the squares $\frac{1}{10}$ $\frac{2}{10}$ $\frac{3}{10}$ $\frac{4}{10}$ $\frac{5}{10}$ $\frac{6}{10}$ $\frac{7}{10}$ $\frac{8}{10}$ $\frac{9}{10}$

Add across →

Add down ↓

Sums are in ◯

$(3^2 - 2^3) \div 10$		0.2
$\dfrac{\sqrt{100} - 1}{10}$		$n > \dfrac{2}{5}$

$\frac{7}{10}$

$1\frac{3}{5}$

$2\frac{1}{5}$

$1\frac{3}{5}$ $1\frac{1}{5}$ $1\frac{7}{10}$

Put these numbers in the squares $\frac{1}{10}$ $\frac{2}{10}$ $\frac{3}{10}$ $\frac{4}{10}$ $\frac{5}{10}$ $\frac{6}{10}$ $\frac{7}{10}$ $\frac{8}{10}$ $\frac{9}{10}$

Add across →

Add down ↓

Sums are in ◯

$\frac{Square}{10}$		$\frac{Odd}{10}$	$1\frac{2}{5}$
	$(0.75 \times 4) \div 5$		$1\frac{2}{5}$
$\frac{Even}{10}$	$(1^2 + 1) \div 10$		$1\frac{7}{10}$
$1\frac{7}{10}$	$1\frac{7}{10}$	$1\frac{1}{10}$	

Put these numbers in the squares $\frac{1}{10}$ $\frac{2}{10}$ $\frac{3}{10}$ $\frac{4}{10}$ $\frac{5}{10}$ $\frac{6}{10}$ $\frac{7}{10}$ $\frac{8}{10}$ $\frac{9}{10}$

Add across $\longrightarrow$

Add down $\downarrow$

Sums are in $\bigcirc$

	$18 \div 60$		$\left(1\frac{2}{5}\right)$
	$\dfrac{\text{Triangular}}{10}$	$1 - \dfrac{1}{10}$	$\left(1\frac{7}{10}\right)$
$\dfrac{4}{5} \div \dfrac{16}{20} \times \dfrac{1}{10}$			$\left(1\frac{2}{5}\right)$

$\left(1\right)$ $\left(1\frac{7}{10}\right)$ $\left(1\frac{4}{5}\right)$

Put these numbers in the squares $\frac{1}{10}$ $\frac{2}{10}$ $\frac{3}{10}$ $\frac{4}{10}$ $\frac{5}{10}$ $\frac{6}{10}$ $\frac{7}{10}$ $\frac{8}{10}$ $\frac{9}{10}$

Add across $\longrightarrow$

Add down $\downarrow$

Sums are in $\bigcirc$

$\frac{1}{2} + \frac{1}{5}$		$1 \div 2$	$1\frac{4}{5}$
$2^3 \div 10$			$1\frac{3}{10}$
		$\dfrac{\text{Prime}}{10}$	$1\frac{2}{5}$
$2\frac{2}{5}$	$\frac{9}{10}$	$1\frac{1}{5}$	

40

Put these numbers in the squares $\frac{1}{10}$ $\frac{2}{10}$ $\frac{3}{10}$ $\frac{4}{10}$ $\frac{5}{10}$ $\frac{6}{10}$ $\frac{7}{10}$ $\frac{8}{10}$ $\frac{9}{10}$

Add across $\longrightarrow$

Add down $\downarrow$

Sums are in $\bigcirc$

		$\frac{\text{Triangular}}{10}$	$1\frac{2}{5}$
	$\frac{|-4| + |+4|}{10}$		$1\frac{9}{10}$
$\frac{\sqrt{1} \times \sqrt{4}}{10} - \frac{1}{10}$		$\frac{18 \div 6 + 6}{10}$	$1\frac{1}{5}$
$1\frac{2}{5}$	$1\frac{1}{2}$	$1\frac{3}{5}$	

41

Put these numbers in the squares $\frac{1}{10}$ $\frac{2}{10}$ $\frac{3}{10}$ $\frac{4}{10}$ $\frac{5}{10}$ $\frac{6}{10}$ $\frac{7}{10}$ $\frac{8}{10}$ $\frac{9}{10}$

Add across $\longrightarrow$

Add down $\downarrow$

Sums are in $\bigcirc$

	$\frac{\text{Odd}}{10}$		$\left(1\frac{2}{5}\right)$
		$8 \div 20$	$\left(1\frac{1}{2}\right)$
	$\sqrt{9} \div 10$	$\frac{\text{Prime}}{10}$	$\left(1\frac{3}{5}\right)$
$\left(1\frac{3}{10}\right)$	$\left(1\frac{3}{10}\right)$	$\left(1\frac{9}{10}\right)$	

Put these numbers in the squares $\frac{1}{10}$ $\frac{2}{10}$ $\frac{3}{10}$ $\frac{4}{10}$ $\frac{5}{10}$ $\frac{6}{10}$ $\frac{7}{10}$ $\frac{8}{10}$ $\frac{9}{10}$

Add across $\longrightarrow$

Add down $\downarrow$

Sums are in $\bigcirc$

$\sqrt{25} \div 10$		
	$\dfrac{\text{Multiple of 9}}{10}$	
$\dfrac{\text{Even} > 7}{10}$		$\dfrac{2 < n < 6}{10}$

$\left(\dfrac{9}{10}\right)$

$\left(1\dfrac{4}{5}\right)$

$\left(1\dfrac{4}{5}\right)$

$\left(1\dfrac{1}{2}\right)$ $\left(1\dfrac{4}{5}\right)$ $\left(1\dfrac{1}{5}\right)$

43

Put these numbers in the squares $\frac{1}{10}$ $\frac{2}{10}$ $\frac{3}{10}$ $\frac{4}{10}$ $\frac{5}{10}$ $\frac{6}{10}$ $\frac{7}{10}$ $\frac{8}{10}$ $\frac{9}{10}$

Add across $\longrightarrow$

Add down $\downarrow$

Sums are in $\bigcirc$

$\dfrac{\text{Triangular}}{10}$		$\sqrt{9} \times \dfrac{1}{10} \times 1$
		$0.1 + 0.2 + 0.3$
$\dfrac{\text{Odd}}{10}$		

$\bigcirc \dfrac{3}{5}$

$\bigcirc 1\dfrac{1}{2}$

$\bigcirc 2\dfrac{2}{5}$

$\bigcirc 1\dfrac{1}{5}$ $\bigcirc 1\dfrac{1}{2}$ $\bigcirc 1\dfrac{4}{5}$

Put these numbers in the squares $\frac{1}{10}$ $\frac{2}{10}$ $\frac{3}{10}$ $\frac{4}{10}$ $\frac{5}{10}$ $\frac{6}{10}$ $\frac{7}{10}$ $\frac{8}{10}$ $\frac{9}{10}$

Add across →

Add down ↓

Sums are in ◯

$\frac{1}{2} < n < 1$	$6^2 \div 6 \div 10$		$1\frac{4}{5}$
			$1\frac{1}{2}$
$0.2 + 0.2 + 0.3$	$2^2 \div 10$		$1\frac{1}{5}$
$2\frac{2}{5}$	$1\frac{1}{2}$	$\frac{3}{5}$	

Put these numbers in the squares $\frac{1}{10}$ $\frac{2}{10}$ $\frac{3}{10}$ $\frac{4}{10}$ $\frac{5}{10}$ $\frac{6}{10}$ $\frac{7}{10}$ $\frac{8}{10}$ $\frac{9}{10}$

Add across →

Add down ↓

Sums are in ◯

$\frac{\text{Multiple of 5}}{10}$			$1\frac{2}{5}$
	$\frac{\text{Odd}}{10}$		$1\frac{3}{5}$
$\sqrt{64} \div 10$		$\frac{57 \times 11 - 626}{10}$	$1\frac{1}{2}$
$1\frac{3}{5}$	$2\frac{1}{5}$	$\frac{7}{10}$	

Put these numbers in the squares $\frac{1}{10}$ $\frac{2}{10}$ $\frac{3}{10}$ $\frac{4}{10}$ $\frac{5}{10}$ $\frac{6}{10}$ $\frac{7}{10}$ $\frac{8}{10}$ $\frac{9}{10}$

Add across $\longrightarrow$

Add down $\downarrow$

Sums are in $\bigcirc$

$\frac{\text{Prime} < 5}{10}$			$\left(1\frac{4}{5}\right)$
		$\frac{\text{Triangular}}{10}$	$\left(1\frac{9}{10}\right)$
$\frac{9}{10} - \frac{4}{5}$		$25 \div 50$	$\left(\frac{4}{5}\right)$
$\left(1\frac{3}{10}\right)$	$\left(1\frac{2}{5}\right)$	$\left(1\frac{4}{5}\right)$	

Put these numbers in the squares $\frac{1}{10}$ $\frac{2}{10}$ $\frac{3}{10}$ $\frac{4}{10}$ $\frac{5}{10}$ $\frac{6}{10}$ $\frac{7}{10}$ $\frac{8}{10}$ $\frac{9}{10}$

Add across ⟶

Add down ↓

Sums are in ◯

	$\frac{100}{100} - \frac{100}{1000}$		$1\frac{2}{5}$
			$1\frac{3}{10}$
$800 \div 1000$		$\frac{600}{1000} - \frac{3}{10}$	$1\frac{4}{5}$
$1\frac{2}{5}$	$1\frac{4}{5}$	$1\frac{3}{10}$	

Put these numbers in the squares $\frac{1}{10}$ $\frac{2}{10}$ $\frac{3}{10}$ $\frac{4}{10}$ $\frac{5}{10}$ $\frac{6}{10}$ $\frac{7}{10}$ $\frac{8}{10}$ $\frac{9}{10}$

Add across →

Add down ↓

Sums are in ◯

$\frac{\sqrt{49}-\sqrt{36}}{10}$	$\frac{\text{Odd}}{10}$		$\frac{4}{5}$
	$4 \div 5$		$1\frac{1}{2}$
		$3^2 \div 10$	$2\frac{1}{5}$
$1\frac{3}{10}$	$1\frac{7}{10}$	$1\frac{1}{2}$	

Put these numbers in the squares $\frac{1}{10}$ $\frac{2}{10}$ $\frac{3}{10}$ $\frac{4}{10}$ $\frac{5}{10}$ $\frac{6}{10}$ $\frac{7}{10}$ $\frac{8}{10}$ $\frac{9}{10}$

Add across $\longrightarrow$

Add down $\downarrow$

Sums are in $\bigcirc$

	$\sqrt{4} \div 10$	
$\dfrac{\lvert -4 \rvert + 5 - 8}{10 \times \lvert -1 \rvert}$		$\sqrt{36} \times 1 \div 10$

$1\frac{3}{5}$

$1\frac{4}{5}$

$1\frac{1}{10}$

$1\frac{7}{10}$　$\frac{9}{10}$　$1\frac{9}{10}$

Put these numbers in the squares $\frac{1}{10}$ $\frac{2}{10}$ $\frac{3}{10}$ $\frac{4}{10}$ $\frac{5}{10}$ $\frac{6}{10}$ $\frac{7}{10}$ $\frac{8}{10}$ $\frac{9}{10}$

Add across ⟶

Add down ↓

Sums are in ◯

		$3 \div 5$	$1\frac{7}{10}$
0.5			$1\frac{1}{2}$
		$4^2 \div 4 \div 10$	$1\frac{3}{10}$

pp

1 $2\frac{2}{5}$ $1\frac{1}{10}$

Put these numbers in the squares $\frac{1}{10}$ $\frac{2}{10}$ $\frac{3}{10}$ $\frac{4}{10}$ $\frac{5}{10}$ $\frac{6}{10}$ $\frac{7}{10}$ $\frac{8}{10}$ $\frac{9}{10}$

Add across →

Add down ↓

Sums are in ◯

$\dfrac{4^2 \div 4 \div 2}{10}$			$\dfrac{9}{10}$
$\dfrac{\sqrt{8^2 + 4^2 + 1}}{10}$		$\dfrac{1^2 \times 1 \times 4 \times 2}{2 \times 5}$	$2\dfrac{1}{5}$
	$\dfrac{2.25 + 0.75}{10}$		$1\dfrac{2}{5}$
$1\dfrac{4}{5}$	$1\dfrac{2}{5}$	$1\dfrac{3}{10}$	

52

Put these numbers in the squares $\frac{1}{10}$ $\frac{2}{10}$ $\frac{3}{10}$ $\frac{4}{10}$ $\frac{5}{10}$ $\frac{6}{10}$ $\frac{7}{10}$ $\frac{8}{10}$ $\frac{9}{10}$

Add across ⟶

Add down ↓

Sums are in ◯

$\frac{\text{Triangular}}{10}$		$\sqrt{4} \div 10$	$\frac{9}{10}$
	$\frac{1}{2} < n < 1$		$2\frac{1}{10}$
0.5		$\frac{\text{Odd}}{10}$	$1\frac{1}{2}$
1	$1\frac{7}{10}$	$1\frac{4}{5}$	

Put these numbers in the squares $\frac{1}{10}$ $\frac{2}{10}$ $\frac{3}{10}$ $\frac{4}{10}$ $\frac{5}{10}$ $\frac{6}{10}$ $\frac{7}{10}$ $\frac{8}{10}$ $\frac{9}{10}$

Add across $\longrightarrow$

Add down $\downarrow$

Sums are in $\bigcirc$

	$\frac{1}{2}+\frac{1}{5}$		$\left(1\frac{1}{5}\right)$
		$3 \div 5$	$\left(1\frac{3}{5}\right)$
$\frac{\text{Prime}}{10}$	$3^2 \div 10$		$\left(1\frac{7}{10}\right)$

$\left(\dfrac{4}{5}\right)$ $\left(2\dfrac{2}{5}\right)$ $\left(1\dfrac{3}{10}\right)$

Put these numbers in the squares $\frac{1}{10}$ $\frac{2}{10}$ $\frac{3}{10}$ $\frac{4}{10}$ $\frac{5}{10}$ $\frac{6}{10}$ $\frac{7}{10}$ $\frac{8}{10}$ $\frac{9}{10}$

Add across →

Add down ↓

Sums are in ◯

$\dfrac{\text{Multiple of 4}}{10}$		$\dfrac{\text{Multiple of 3}}{10}$
	$2^2 \div 10$	
$3 \div 10$		

Circles right of grid (top to bottom):

$1\frac{9}{10}$

$\frac{7}{10}$

$1\frac{9}{10}$

Circles below grid (left to right):

$1\frac{1}{5}$ $1\frac{4}{5}$ $1\frac{1}{2}$

55

Put these numbers in the squares $\frac{1}{10}$ $\frac{2}{10}$ $\frac{3}{10}$ $\frac{4}{10}$ $\frac{5}{10}$ $\frac{6}{10}$ $\frac{7}{10}$ $\frac{8}{10}$ $\frac{9}{10}$

Add across $\longrightarrow$

Add down $\downarrow$

Sums are in $\bigcirc$

		$\dfrac{\sqrt{100} \div \sqrt{100}}{10}$
	$\dfrac{\text{Even Prime}}{10}$	
	$30 \div 6 \div 10$	$\dfrac{\text{Prime}}{10}$

$1\frac{7}{10}$

$1\frac{1}{5}$

$1\frac{3}{5}$

$2\frac{1}{10}$ $1\frac{2}{5}$ 1

Put these numbers in the squares $\frac{1}{10}$ $\frac{2}{10}$ $\frac{3}{10}$ $\frac{4}{10}$ $\frac{5}{10}$ $\frac{6}{10}$ $\frac{7}{10}$ $\frac{8}{10}$ $\frac{9}{10}$

Add across $\longrightarrow$

Add down $\downarrow$

Sums are in $\bigcirc$

		$\sqrt{81} \div 10$
$6 \div 10$	0.5	

Circles (right): $1\frac{2}{5}$, $1\frac{3}{10}$, $1\frac{4}{5}$

Circles (bottom): $1\frac{3}{10}$, $1\frac{2}{5}$, $1\frac{4}{5}$

Put these numbers in the squares $\frac{1}{10}$ $\frac{2}{10}$ $\frac{3}{10}$ $\frac{4}{10}$ $\frac{5}{10}$ $\frac{6}{10}$ $\frac{7}{10}$ $\frac{8}{10}$ $\frac{9}{10}$

Add across $\longrightarrow$

Add down $\downarrow$

Sums are in $\bigcirc$

$1.5 \times 2 \div 5$		
		$n > \dfrac{2}{5}$
		$\dfrac{\sqrt{90} \times 0 + \sqrt{9}}{10}$

$\left(2\right)$

$\left(1\dfrac{1}{5}\right)$

$\left(1\dfrac{3}{10}\right)$

$\left(1\dfrac{1}{2}\right)$ $\left(1\dfrac{1}{10}\right)$ $\left(1\dfrac{9}{10}\right)$

Put these numbers in the squares $\frac{1}{10}$ $\frac{2}{10}$ $\frac{3}{10}$ $\frac{4}{10}$ $\frac{5}{10}$ $\frac{6}{10}$ $\frac{7}{10}$ $\frac{8}{10}$ $\frac{9}{10}$

Add across →

Add down ↓

Sums are in ◯

$\sqrt{100} \times \frac{1}{10} \div 10$	$\frac{Square}{10}$		$1\frac{4}{5}$
$\frac{Triangular}{10}$			$1\frac{1}{2}$
		$\frac{1}{10} + \frac{1}{5}$	$1\frac{1}{5}$
$1\frac{1}{5}$	$1\frac{1}{2}$	$1\frac{4}{5}$	

Put these numbers in the squares $\frac{1}{10}$ $\frac{2}{10}$ $\frac{3}{10}$ $\frac{4}{10}$ $\frac{5}{10}$ $\frac{6}{10}$ $\frac{7}{10}$ $\frac{8}{10}$ $\frac{9}{10}$

Add across $\longrightarrow$

Add down $\downarrow$

Sums are in $\bigcirc$

	$n > \frac{2}{5}$	$3^2 \div 10$	$2\frac{3}{10}$
	$\sqrt{100} \times \frac{1}{10} \div 10$		$\frac{3}{5}$
$1 \div 2$			$1\frac{3}{5}$
$1\frac{2}{5}$	$1\frac{3}{10}$	$1\frac{4}{5}$	

Put these numbers in the squares $\frac{1}{10}$ $\frac{2}{10}$ $\frac{3}{10}$ $\frac{4}{10}$ $\frac{5}{10}$ $\frac{6}{10}$ $\frac{7}{10}$ $\frac{8}{10}$ $\frac{9}{10}$

Add across ⟶

Add down ↓

Sums are in ◯

$\frac{\text{Square}}{10}$		
	$\frac{\text{Prime}}{10}$	
$\frac{2}{4} + \frac{1}{5}$		$0.09 + 0.01$

$2\frac{3}{10}$

$\frac{9}{10}$

$1\frac{3}{10}$

$1\frac{9}{10}$ $1\frac{3}{10}$ $1\frac{3}{10}$

61

Put these numbers in the squares $\frac{1}{10}$ $\frac{2}{10}$ $\frac{3}{10}$ $\frac{4}{10}$ $\frac{5}{10}$ $\frac{6}{10}$ $\frac{7}{10}$ $\frac{8}{10}$ $\frac{9}{10}$

Add across $\longrightarrow$

Add down $\downarrow$

Sums are in $\bigcirc$

	$\dfrac{\text{Triangular}}{10}$	$\sqrt{36} \div 10$	$\left(1\frac{4}{5}\right)$
			$\left(1\frac{1}{2}\right)$
$\sqrt{49} \div 10$			$\left(1\frac{1}{5}\right)$
$\left(2\frac{2}{5}\right)$	$\left(\frac{3}{5}\right)$	$\left(1\frac{1}{2}\right)$	

Put these numbers in the squares $\frac{1}{10}$ $\frac{2}{10}$ $\frac{3}{10}$ $\frac{4}{10}$ $\frac{5}{10}$ $\frac{6}{10}$ $\frac{7}{10}$ $\frac{8}{10}$ $\frac{9}{10}$

Add across $\longrightarrow$

Add down $\downarrow$

Sums are in $\bigcirc$

	$\sqrt{1} \div 10$		
			$\frac{4}{5}$
$\sqrt{16} \div 10$			
			$1\frac{2}{5}$
	$\dfrac{\text{Square}}{10}$		
			$2\frac{3}{10}$

$1\frac{1}{5}$ $1\frac{3}{10}$ 2

63

Put these numbers in the squares $\frac{1}{10}$ $\frac{2}{10}$ $\frac{3}{10}$ $\frac{4}{10}$ $\frac{5}{10}$ $\frac{6}{10}$ $\frac{7}{10}$ $\frac{8}{10}$ $\frac{9}{10}$

Add across $\longrightarrow$

Add down $\downarrow$

Sums are in $\bigcirc$

$8 \times \frac{1}{10}$		$\frac{\text{Triangular}}{10}$
	$6 \div 10$	
		$\frac{\text{Prime}}{10}$

$1\frac{1}{5}$

$1\frac{9}{10}$

$1\frac{2}{5}$

$1\frac{2}{5}$ $1\frac{3}{5}$ $1\frac{1}{2}$

64

Put these numbers in the squares $\frac{1}{10}$ $\frac{2}{10}$ $\frac{3}{10}$ $\frac{4}{10}$ $\frac{5}{10}$ $\frac{6}{10}$ $\frac{7}{10}$ $\frac{8}{10}$ $\frac{9}{10}$

Add across $\longrightarrow$

Add down $\downarrow$

Sums are in $\bigcirc$

$\frac{\text{Square}}{10}$			$1\frac{2}{5}$
$1-\frac{7}{10}$		$\frac{\text{Square}}{10}$	1
$\frac{n>6}{10}$		$n<0.6$	$2\frac{1}{10}$
$1\frac{3}{5}$	$1\frac{1}{2}$	$1\frac{2}{5}$	

Put these numbers in the squares $\frac{1}{10}$ $\frac{2}{10}$ $\frac{3}{10}$ $\frac{4}{10}$ $\frac{5}{10}$ $\frac{6}{10}$ $\frac{7}{10}$ $\frac{8}{10}$ $\frac{9}{10}$

Add across →

Add down ↓

Sums are in ◯

		$n > \frac{2}{5}$	$1\frac{1}{5}$
	$\left\|-\frac{2}{20}\right\|$	$\frac{\text{Odd}}{10}$	$1\frac{2}{5}$
$4 \div 5$			$1\frac{9}{10}$
$1\frac{7}{10}$	$1\frac{2}{5}$	$1\frac{2}{5}$	

66

Put these numbers in the squares $\frac{1}{10}$ $\frac{2}{10}$ $\frac{3}{10}$ $\frac{4}{10}$ $\frac{5}{10}$ $\frac{6}{10}$ $\frac{7}{10}$ $\frac{8}{10}$ $\frac{9}{10}$

Add across ➝

Add down ↓

Sums are in ◯

			$1\frac{1}{5}$
$\dfrac{\text{Square}}{10}$	$n > \dfrac{2}{5}$	$n < \dfrac{1}{2}$	$1\frac{7}{10}$
$1 \div 2$			$1\frac{3}{5}$
$2\frac{1}{10}$	$1\frac{3}{10}$	$1\frac{1}{10}$	

Put these numbers in the squares $\frac{1}{10}$ $\frac{2}{10}$ $\frac{3}{10}$ $\frac{4}{10}$ $\frac{5}{10}$ $\frac{6}{10}$ $\frac{7}{10}$ $\frac{8}{10}$ $\frac{9}{10}$

Add across ⟶

Add down ↓

Sums are in ◯

		$\frac{1}{10} \div \frac{1}{5} + \frac{1}{5}$	$1\frac{1}{2}$
	$\frac{Square}{10}$		$1\frac{2}{5}$
	$\frac{3}{2} \times \frac{3}{5}$		$1\frac{3}{5}$
$1\frac{4}{5}$	$1\frac{1}{5}$	$1\frac{1}{2}$	

68

Put these numbers in the squares $\frac{1}{10}$ $\frac{2}{10}$ $\frac{3}{10}$ $\frac{4}{10}$ $\frac{5}{10}$ $\frac{6}{10}$ $\frac{7}{10}$ $\frac{8}{10}$ $\frac{9}{10}$

Add across $\longrightarrow$

Add down $\downarrow$

Sums are in $\bigcirc$

		$1 - \dfrac{1}{10}$
$\dfrac{\lvert -1 \rvert + 1 - 1}{10}$		
		$\dfrac{\lvert -1 \rvert + 2}{10}$

Right circles: $2\frac{1}{5}$, $1\frac{3}{10}$, 1

Bottom circles: 1, $1\frac{9}{10}$, $1\frac{3}{5}$

Put these numbers in the squares $\frac{1}{10}$ $\frac{2}{10}$ $\frac{3}{10}$ $\frac{4}{10}$ $\frac{5}{10}$ $\frac{6}{10}$ $\frac{7}{10}$ $\frac{8}{10}$ $\frac{9}{10}$

Add across ⟶

Add down ↓

Sums are in ◯

			$2\frac{3}{10}$
$\frac{\sqrt{100}}{10} - \frac{9}{10}$	$32 \div 16 \div 10$		$\frac{7}{10}$
		$\frac{\lvert 2 - 7 \rvert}{10}$	$1\frac{1}{2}$
$1\frac{3}{10}$	$1\frac{7}{10}$	$1\frac{1}{2}$	

Put these numbers in the squares $\frac{1}{10}$ $\frac{2}{10}$ $\frac{3}{10}$ $\frac{4}{10}$ $\frac{5}{10}$ $\frac{6}{10}$ $\frac{7}{10}$ $\frac{8}{10}$ $\frac{9}{10}$

Add across →

Add down ↓

Sums are in ◯

		0.5	$\frac{9}{10}$
	$\dfrac{4 \times 21 \div 7 \div 3}{10}$		$1\frac{4}{5}$
$\dfrac{80}{400}$	$\dfrac{\text{Prime}}{10}$		$1\frac{4}{5}$
$1\frac{1}{10}$	$1\frac{2}{5}$	2	

Put these numbers in the squares $\frac{1}{10}$ $\frac{2}{10}$ $\frac{3}{10}$ $\frac{4}{10}$ $\frac{5}{10}$ $\frac{6}{10}$ $\frac{7}{10}$ $\frac{8}{10}$ $\frac{9}{10}$

Add across $\longrightarrow$

Add down $\downarrow$

Sums are in $\bigcirc$

$\dfrac{\left(\sqrt{81}-31\right)^0}{10}$			$1\frac{1}{2}$
	$\lvert-3\rvert \div 10$		$1\frac{1}{2}$
$\dfrac{\text{Odd} < 8}{10}$		$\dfrac{40 \div 40 \text{ x } \sqrt{4}}{10}$	$1\frac{1}{2}$
$1\frac{1}{5}$	$1\frac{4}{5}$	$1\frac{1}{2}$	

Put these numbers in the squares $\frac{1}{10}$ $\frac{2}{10}$ $\frac{3}{10}$ $\frac{4}{10}$ $\frac{5}{10}$ $\frac{6}{10}$ $\frac{7}{10}$ $\frac{8}{10}$ $\frac{9}{10}$

Add across →

Add down ↓

Sums are in ◯

$\frac{\text{Even Triangular}}{10}$			$1\frac{3}{5}$
	0.9	$\frac{\text{Even Prime}}{10}$	$1\frac{3}{5}$
$\frac{\text{Square}}{10}$			$1\frac{3}{10}$
$1\frac{1}{2}$	$1\frac{7}{10}$	$1\frac{3}{10}$	

Put these numbers in the squares $\frac{1}{10}$ $\frac{2}{10}$ $\frac{3}{10}$ $\frac{4}{10}$ $\frac{5}{10}$ $\frac{6}{10}$ $\frac{7}{10}$ $\frac{8}{10}$ $\frac{9}{10}$

Add across $\longrightarrow$

Add down $\downarrow$

Sums are in $\bigcirc$

		$3^2 \div 10$	$1\frac{7}{10}$
$\frac{\text{Prime}}{10}$			$1\frac{1}{2}$
	$6 \div 10$	$\frac{\text{Square}}{10}$	$1\frac{3}{10}$
$\frac{9}{10}$	$1\frac{1}{2}$	$2\frac{1}{10}$	

74

Put these numbers in the squares $\frac{1}{10}$ $\frac{2}{10}$ $\frac{3}{10}$ $\frac{4}{10}$ $\frac{5}{10}$ $\frac{6}{10}$ $\frac{7}{10}$ $\frac{8}{10}$ $\frac{9}{10}$

Add across $\longrightarrow$

Add down $\downarrow$

Sums are in $\bigcirc$

$1 \div 2$			$1\frac{1}{10}$
	$\dfrac{58 \times 11 - 637}{10}$		$1\frac{2}{5}$
		$\dfrac{\text{Triangular}}{10}$	2
$2\frac{1}{10}$	$1\frac{3}{10}$	$1\frac{1}{10}$	

Put these numbers in the squares $\frac{1}{10}$ $\frac{2}{10}$ $\frac{3}{10}$ $\frac{4}{10}$ $\frac{5}{10}$ $\frac{6}{10}$ $\frac{7}{10}$ $\frac{8}{10}$ $\frac{9}{10}$

Add across $\longrightarrow$

Add down $\downarrow$

Sums are in $\bigcirc$

	$\frac{Prime}{10}$	$\frac{Prime}{10}$	$1\frac{1}{10}$
$0.3 + \frac{1}{5}$	$n < \frac{2}{5}$		$1\frac{1}{10}$
		$\frac{Multiple\ of\ 3}{10}$	$2\frac{3}{10}$
$1\frac{4}{10}$	$1\frac{1}{10}$	2	

Put these numbers in the squares $\frac{1}{10}$ $\frac{2}{10}$ $\frac{3}{10}$ $\frac{4}{10}$ $\frac{5}{10}$ $\frac{6}{10}$ $\frac{7}{10}$ $\frac{8}{10}$ $\frac{9}{10}$

Add across $\longrightarrow$

Add down $\downarrow$

Sums are in $\bigcirc$

$\dfrac{\text{Prime}}{10}$		
	$\sqrt{16} \div 10$	
$\sqrt{81} \div 10$		

$1\dfrac{1}{2}$

$1\dfrac{1}{5}$

$1\dfrac{4}{5}$

$2\dfrac{1}{5}$ $1\dfrac{7}{10}$ $\dfrac{6}{10}$

Put these numbers in the squares $\frac{1}{10}$ $\frac{2}{10}$ $\frac{3}{10}$ $\frac{4}{10}$ $\frac{5}{10}$ $\frac{6}{10}$ $\frac{7}{10}$ $\frac{8}{10}$ $\frac{9}{10}$

Add across ⟶

Add down ↓

Sums are in ◯

	$\frac{\text{Triangular}}{10}$		◯ 1
	$\frac{\text{Prime}}{10}$		◯ $1\frac{2}{5}$
	$\frac{2}{10} + \frac{3}{5}$	$\frac{\text{Multiple of 3}}{10}$	◯ $2\frac{1}{10}$

◯ $\frac{7}{10}$ ◯ $1\frac{3}{5}$ ◯ $2\frac{1}{5}$

Put these numbers in the squares $\frac{1}{10}$ $\frac{2}{10}$ $\frac{3}{10}$ $\frac{4}{10}$ $\frac{5}{10}$ $\frac{6}{10}$ $\frac{7}{10}$ $\frac{8}{10}$ $\frac{9}{10}$

Add across $\longrightarrow$

Add down $\downarrow$

Sums are in $\bigcirc$

$\dfrac{\text{Square}}{10}$	$\sqrt{16} \div 10$		$1\dfrac{1}{5}$
	$\dfrac{\text{Even Prime}}{10}$	$\dfrac{\text{Prime}}{10}$	1
			$2\dfrac{3}{10}$
$1\dfrac{1}{2}$	$1\dfrac{1}{5}$	$1\dfrac{4}{5}$	

1.

$\frac{4}{10}$	$\frac{9}{10}$	$\frac{1}{10}$	$1\frac{2}{5}$
$\frac{7}{10}$	$\frac{3}{10}$	$\frac{8}{10}$	$1\frac{4}{5}$
$\frac{2}{10}$	$\frac{5}{10}$	$\frac{6}{10}$	$1\frac{3}{10}$
$1\frac{3}{10}$	$1\frac{7}{10}$	$1\frac{1}{2}$	

2.

$\frac{9}{10}$	$\frac{6}{10}$	$\frac{1}{10}$	$1\frac{3}{5}$
$\frac{3}{10}$	$\frac{5}{10}$	$\frac{8}{10}$	$1\frac{3}{5}$
$\frac{2}{10}$	$\frac{4}{10}$	$\frac{7}{10}$	$1\frac{3}{10}$
$1\frac{2}{5}$	$1\frac{1}{2}$	$1\frac{3}{5}$	

3.

$\frac{9}{10}$	$\frac{6}{10}$	$\frac{7}{10}$	$2\frac{1}{5}$
$\frac{5}{10}$	$\frac{1}{10}$	$\frac{2}{10}$	$\frac{4}{5}$
$\frac{4}{10}$	$\frac{3}{10}$	$\frac{8}{10}$	$1\frac{1}{2}$
$1\frac{4}{5}$	1	$1\frac{7}{10}$	

4.

$\frac{4}{10}$	$\frac{8}{10}$	$\frac{9}{10}$	$2\frac{1}{10}$
$\frac{1}{10}$	$\frac{6}{10}$	$\frac{7}{10}$	$1\frac{2}{5}$
$\frac{2}{10}$	$\frac{3}{10}$	$\frac{5}{10}$	1
$\frac{7}{10}$	$1\frac{7}{10}$	$2\frac{1}{10}$	

5.

$\frac{7}{10}$	$\frac{6}{10}$	$\frac{9}{10}$	$2\frac{1}{5}$
$\frac{1}{10}$	$\frac{2}{10}$	$\frac{8}{10}$	$1\frac{1}{10}$
$\frac{5}{10}$	$\frac{4}{10}$	$\frac{3}{10}$	$1\frac{1}{5}$
$1\frac{3}{10}$	$1\frac{1}{5}$	2	

6.

$\frac{4}{10}$	$\frac{9}{10}$	$\frac{6}{10}$	$1\frac{9}{10}$
$\frac{8}{10}$	$\frac{1}{10}$	$\frac{5}{10}$	$1\frac{2}{5}$
$\frac{3}{10}$	$\frac{7}{10}$	$\frac{2}{10}$	$1\frac{1}{5}$
$1\frac{1}{2}$	$1\frac{7}{10}$	$1\frac{3}{10}$	

7.

$\frac{4}{10}$	$\frac{6}{10}$	$\frac{2}{10}$	$1\frac{1}{5}$
$\frac{1}{10}$	$\frac{5}{10}$	$\frac{3}{10}$	$\frac{9}{10}$
$\frac{9}{10}$	$\frac{7}{10}$	$\frac{8}{10}$	$2\frac{2}{5}$
$1\frac{2}{5}$	$1\frac{4}{5}$	$1\frac{3}{10}$	

8.

$\frac{7}{10}$	$\frac{2}{10}$	$\frac{6}{10}$	$1\frac{1}{2}$
$\frac{4}{10}$	$\frac{5}{10}$	$\frac{8}{10}$	$1\frac{7}{10}$
$\frac{1}{10}$	$\frac{3}{10}$	$\frac{9}{10}$	$1\frac{3}{10}$
$1\frac{1}{5}$	1	$2\frac{3}{10}$	

9.

$\frac{8}{10}$	$\frac{3}{10}$	$\frac{7}{10}$	$1\frac{4}{5}$
$\frac{6}{10}$	$\frac{9}{10}$	$\frac{1}{10}$	$1\frac{3}{5}$
$\frac{4}{10}$	$\frac{2}{10}$	$\frac{5}{10}$	$1\frac{1}{10}$
$1\frac{4}{5}$	$1\frac{2}{5}$	$1\frac{3}{10}$	

10.

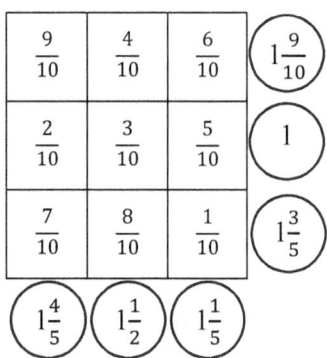

$\frac{9}{10}$	$\frac{4}{10}$	$\frac{6}{10}$	$1\frac{9}{10}$
$\frac{2}{10}$	$\frac{3}{10}$	$\frac{5}{10}$	1
$\frac{7}{10}$	$\frac{8}{10}$	$\frac{1}{10}$	$1\frac{3}{5}$
$1\frac{4}{5}$	$1\frac{1}{2}$	$1\frac{1}{5}$	

11.

$\frac{3}{10}$	$\frac{6}{10}$	$\frac{2}{10}$	$1\frac{1}{10}$
$\frac{8}{10}$	$\frac{9}{10}$	$\frac{4}{10}$	$2\frac{1}{10}$
$\frac{5}{10}$	$\frac{1}{10}$	$\frac{7}{10}$	$1\frac{3}{10}$
$1\frac{3}{5}$	$1\frac{3}{5}$	$1\frac{3}{10}$	

12.

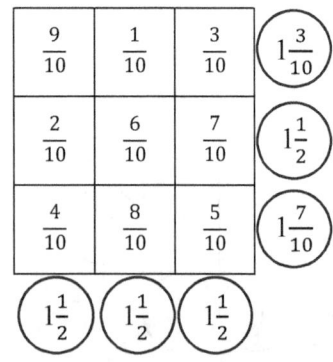

$\frac{9}{10}$	$\frac{1}{10}$	$\frac{3}{10}$	$1\frac{3}{10}$
$\frac{2}{10}$	$\frac{6}{10}$	$\frac{7}{10}$	$1\frac{1}{2}$
$\frac{4}{10}$	$\frac{8}{10}$	$\frac{5}{10}$	$1\frac{7}{10}$
$1\frac{1}{2}$	$1\frac{1}{2}$	$1\frac{1}{2}$	

13.

$\frac{4}{10}$	$\frac{9}{10}$	$\frac{5}{10}$	$1\frac{4}{5}$
$\frac{8}{10}$	$\frac{1}{10}$	$\frac{6}{10}$	$1\frac{1}{2}$
$\frac{3}{10}$	$\frac{7}{10}$	$\frac{2}{10}$	$1\frac{1}{5}$
$1\frac{1}{2}$	$1\frac{7}{10}$	$1\frac{3}{10}$	

14.

$\frac{7}{10}$	$\frac{5}{10}$	$\frac{3}{10}$	$1\frac{1}{2}$
$\frac{4}{10}$	$\frac{1}{10}$	$\frac{8}{10}$	$1\frac{3}{10}$
$\frac{9}{10}$	$\frac{2}{10}$	$\frac{6}{10}$	$1\frac{7}{10}$
2	$\frac{4}{5}$	$1\frac{7}{10}$	

15.

$\frac{3}{10}$	$\frac{2}{10}$	$\frac{9}{10}$	$1\frac{2}{5}$
$\frac{4}{10}$	$\frac{1}{10}$	$\frac{8}{10}$	$1\frac{3}{10}$
$\frac{5}{10}$	$\frac{6}{10}$	$\frac{7}{10}$	$1\frac{4}{5}$
$1\frac{1}{5}$	$\frac{9}{10}$	$2\frac{2}{5}$	

16.

$\frac{7}{10}$	$\frac{2}{10}$	$\frac{5}{10}$	$1\frac{2}{5}$
$\frac{6}{10}$	$\frac{9}{10}$	$\frac{8}{10}$	$2\frac{3}{10}$
$\frac{3}{10}$	$\frac{4}{10}$	$\frac{1}{10}$	$\frac{4}{5}$
$1\frac{3}{5}$	$1\frac{1}{2}$	$1\frac{2}{5}$	

17.

$\frac{8}{10}$	$\frac{5}{10}$	$\frac{7}{10}$	2
$\frac{3}{10}$	$\frac{1}{10}$	$\frac{2}{10}$	$\frac{3}{5}$
$\frac{6}{10}$	$\frac{9}{10}$	$\frac{4}{10}$	$1\frac{9}{10}$
$1\frac{7}{10}$	$1\frac{1}{2}$	$1\frac{3}{10}$	

18.

$\frac{6}{10}$	$\frac{8}{10}$	$\frac{1}{10}$	$1\frac{1}{2}$
$\frac{3}{10}$	$\frac{2}{10}$	$\frac{5}{10}$	1
$\frac{7}{10}$	$\frac{4}{10}$	$\frac{9}{10}$	2
$1\frac{3}{5}$	$1\frac{2}{5}$	$1\frac{1}{2}$	

19.

$\frac{8}{10}$	$\frac{5}{10}$	$\frac{6}{10}$	$1\frac{9}{10}$
$\frac{3}{10}$	$\frac{9}{10}$	$\frac{4}{10}$	$1\frac{3}{5}$
$\frac{1}{10}$	$\frac{7}{10}$	$\frac{2}{10}$	1
$1\frac{1}{5}$	$2\frac{1}{10}$	$1\frac{1}{5}$	

20.

$\frac{1}{10}$	$\frac{2}{10}$	$\frac{9}{10}$	$1\frac{1}{5}$
$\frac{8}{10}$	$\frac{5}{10}$	$\frac{6}{10}$	$1\frac{9}{10}$
$\frac{4}{10}$	$\frac{7}{10}$	$\frac{3}{10}$	$1\frac{2}{5}$
$1\frac{3}{10}$	$1\frac{2}{5}$	$1\frac{4}{5}$	

21.

$\frac{2}{10}$	$\frac{8}{10}$	$\frac{5}{10}$	$1\frac{1}{2}$
$\frac{7}{10}$	$\frac{4}{10}$	$\frac{9}{10}$	2
$\frac{6}{10}$	$\frac{1}{10}$	$\frac{3}{10}$	1
$1\frac{1}{2}$	$1\frac{3}{10}$	$1\frac{7}{10}$	

22.

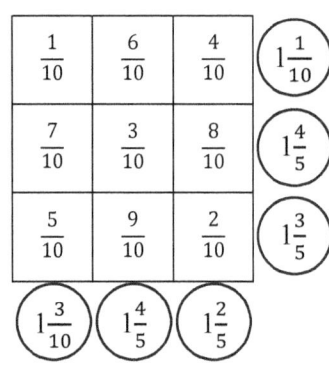

$\frac{1}{10}$	$\frac{6}{10}$	$\frac{4}{10}$	$1\frac{1}{10}$
$\frac{7}{10}$	$\frac{3}{10}$	$\frac{8}{10}$	$1\frac{4}{5}$
$\frac{5}{10}$	$\frac{9}{10}$	$\frac{2}{10}$	$1\frac{3}{5}$
$1\frac{3}{10}$	$1\frac{4}{5}$	$1\frac{2}{5}$	

23.

$\frac{8}{10}$	$\frac{7}{10}$	$\frac{6}{10}$	$2\frac{1}{10}$
$\frac{4}{10}$	$\frac{2}{10}$	$\frac{9}{10}$	$1\frac{1}{2}$
$\frac{5}{10}$	$\frac{3}{10}$	$\frac{1}{10}$	$\frac{9}{10}$
$1\frac{7}{10}$	$1\frac{1}{5}$	$1\frac{3}{5}$	

24.

$\frac{1}{10}$	$\frac{9}{10}$	$\frac{7}{10}$	$1\frac{7}{10}$
$\frac{6}{10}$	$\frac{5}{10}$	$\frac{4}{10}$	$1\frac{1}{2}$
$\frac{3}{10}$	$\frac{2}{10}$	$\frac{8}{10}$	$1\frac{3}{10}$
1	$1\frac{3}{5}$	$1\frac{9}{10}$	

25.

$\frac{2}{10}$	$\frac{9}{10}$	$\frac{8}{10}$	$1\frac{9}{10}$
$\frac{6}{10}$	$\frac{4}{10}$	$\frac{1}{10}$	$1\frac{1}{10}$
$\frac{3}{10}$	$\frac{7}{10}$	$\frac{5}{10}$	$1\frac{1}{2}$
$1\frac{1}{10}$	2	$1\frac{2}{5}$	

26.

$\frac{4}{10}$	$\frac{2}{10}$	$\frac{7}{10}$	$1\frac{3}{10}$
$\frac{6}{10}$	$\frac{5}{10}$	$\frac{8}{10}$	$1\frac{9}{10}$
$\frac{1}{10}$	$\frac{3}{10}$	$\frac{9}{10}$	$1\frac{3}{10}$
$1\frac{1}{10}$	1	$2\frac{2}{5}$	

27.

$\frac{8}{10}$	$\frac{1}{10}$	$\frac{9}{10}$	$1\frac{4}{5}$
$\frac{3}{10}$	$\frac{7}{10}$	$\frac{4}{10}$	$1\frac{2}{5}$
$\frac{6}{10}$	$\frac{2}{10}$	$\frac{5}{10}$	$1\frac{3}{10}$
$1\frac{7}{10}$	1	$1\frac{4}{5}$	

28.

$\frac{1}{10}$	$\frac{3}{10}$	$\frac{5}{10}$	$\frac{9}{10}$
$\frac{8}{10}$	$\frac{2}{10}$	$\frac{6}{10}$	$1\frac{3}{5}$
$\frac{9}{10}$	$\frac{4}{10}$	$\frac{7}{10}$	2
$1\frac{4}{5}$	$\frac{9}{10}$	$1\frac{4}{5}$	

29.

$\frac{3}{10}$	$\frac{9}{10}$	$\frac{5}{10}$	$1\frac{7}{10}$
$\frac{1}{10}$	$\frac{7}{10}$	$\frac{2}{10}$	1
$\frac{8}{10}$	$\frac{4}{10}$	$\frac{6}{10}$	$1\frac{4}{5}$
$1\frac{1}{5}$	2	$1\frac{3}{10}$	

30.

$\frac{4}{10}$	$\frac{3}{10}$	$\frac{8}{10}$	$1\frac{1}{2}$
$\frac{6}{10}$	$\frac{1}{10}$	$\frac{5}{10}$	$1\frac{1}{5}$
$\frac{2}{10}$	$\frac{9}{10}$	$\frac{7}{10}$	$1\frac{4}{5}$
$1\frac{1}{5}$	$1\frac{3}{10}$	2	

31.

$\frac{2}{10}$	$\frac{7}{10}$	$\frac{9}{10}$	$1\frac{4}{5}$
$\frac{5}{10}$	$\frac{3}{10}$	$\frac{4}{10}$	$1\frac{1}{5}$
$\frac{1}{10}$	$\frac{6}{10}$	$\frac{8}{10}$	$1\frac{1}{2}$

$\frac{4}{5}$ $1\frac{3}{5}$ $2\frac{1}{10}$

32.

$\frac{2}{10}$	$\frac{9}{10}$	$\frac{5}{10}$	$1\frac{3}{5}$
$\frac{6}{10}$	$\frac{1}{10}$	$\frac{8}{10}$	$1\frac{1}{2}$
$\frac{4}{10}$	$\frac{7}{10}$	$\frac{3}{10}$	$1\frac{2}{5}$

$1\frac{1}{5}$ $1\frac{7}{10}$ $1\frac{3}{5}$

33.

$\frac{8}{10}$	$\frac{1}{10}$	$\frac{5}{10}$	$1\frac{2}{5}$
$\frac{2}{10}$	$\frac{9}{10}$	$\frac{3}{10}$	$1\frac{2}{5}$
$\frac{7}{10}$	$\frac{4}{10}$	$\frac{6}{10}$	$1\frac{7}{10}$

$1\frac{7}{10}$ $1\frac{2}{5}$ $1\frac{2}{5}$

34.

$\frac{9}{10}$	$\frac{7}{10}$	$\frac{4}{10}$	2
$\frac{3}{10}$	$\frac{1}{10}$	$\frac{2}{10}$	$\frac{3}{5}$
$\frac{6}{10}$	$\frac{8}{10}$	$\frac{5}{10}$	$1\frac{9}{10}$

$1\frac{4}{5}$ $1\frac{3}{5}$ $1\frac{1}{10}$

35.

$\frac{3}{10}$	$\frac{8}{10}$	$\frac{6}{10}$	$1\frac{7}{10}$
$\frac{7}{10}$	$\frac{9}{10}$	$\frac{5}{10}$	$2\frac{1}{10}$
$\frac{2}{10}$	$\frac{4}{10}$	$\frac{1}{10}$	$\frac{7}{10}$

$1\frac{1}{5}$ $2\frac{1}{10}$ $1\frac{1}{5}$

36.

$\frac{1}{10}$	$\frac{4}{10}$	$\frac{2}{10}$	$\frac{7}{10}$
$\frac{6}{10}$	$\frac{3}{10}$	$\frac{7}{10}$	$1\frac{3}{5}$
$\frac{9}{10}$	$\frac{5}{10}$	$\frac{8}{10}$	$2\frac{1}{5}$

$1\frac{3}{5}$ $1\frac{1}{5}$ $1\frac{7}{10}$

37.

$\frac{4}{10}$	$\frac{9}{10}$	$\frac{1}{10}$	$1\frac{2}{5}$
$\frac{5}{10}$	$\frac{6}{10}$	$\frac{3}{10}$	$1\frac{2}{5}$
$\frac{8}{10}$	$\frac{2}{10}$	$\frac{7}{10}$	$1\frac{7}{10}$
$1\frac{7}{10}$	$1\frac{7}{10}$	$1\frac{1}{10}$	

38.

$\frac{7}{10}$	$\frac{3}{10}$	$\frac{4}{10}$	$1\frac{2}{5}$
$\frac{2}{10}$	$\frac{6}{10}$	$\frac{9}{10}$	$1\frac{7}{10}$
$\frac{1}{10}$	$\frac{8}{10}$	$\frac{5}{10}$	$1\frac{2}{5}$
1	$1\frac{7}{10}$	$1\frac{4}{5}$	

39.

$\frac{7}{10}$	$\frac{6}{10}$	$\frac{5}{10}$	$1\frac{4}{5}$
$\frac{8}{10}$	$\frac{1}{10}$	$\frac{4}{10}$	$1\frac{3}{10}$
$\frac{9}{10}$	$\frac{2}{10}$	$\frac{3}{10}$	$1\frac{2}{5}$
$2\frac{2}{5}$	$\frac{9}{10}$	$1\frac{1}{5}$	

40.

$\frac{6}{10}$	$\frac{5}{10}$	$\frac{3}{10}$	$1\frac{2}{5}$
$\frac{7}{10}$	$\frac{8}{10}$	$\frac{4}{10}$	$1\frac{9}{10}$
$\frac{1}{10}$	$\frac{2}{10}$	$\frac{9}{10}$	$1\frac{1}{5}$
$1\frac{2}{5}$	$1\frac{1}{2}$	$1\frac{3}{5}$	

41.

$\frac{5}{10}$	$\frac{1}{10}$	$\frac{8}{10}$	$1\frac{2}{5}$
$\frac{2}{10}$	$\frac{9}{10}$	$\frac{4}{10}$	$1\frac{1}{2}$
$\frac{6}{10}$	$\frac{3}{10}$	$\frac{7}{10}$	$1\frac{3}{5}$
$1\frac{3}{10}$	$1\frac{3}{10}$	$1\frac{9}{10}$	

42.

$\frac{5}{10}$	$\frac{3}{10}$	$\frac{1}{10}$	$\frac{9}{10}$
$\frac{2}{10}$	$\frac{9}{10}$	$\frac{7}{10}$	$1\frac{4}{5}$
$\frac{8}{10}$	$\frac{6}{10}$	$\frac{4}{10}$	$1\frac{4}{5}$
$1\frac{1}{2}$	$1\frac{4}{5}$	$1\frac{1}{5}$	

43.

$\frac{1}{10}$	$\frac{2}{10}$	$\frac{3}{10}$	$\frac{3}{5}$
$\frac{4}{10}$	$\frac{5}{10}$	$\frac{6}{10}$	$1\frac{1}{2}$
$\frac{7}{10}$	$\frac{8}{10}$	$\frac{9}{10}$	$2\frac{2}{5}$
$1\frac{1}{5}$	$1\frac{1}{2}$	$1\frac{4}{5}$	

44.

$\frac{9}{10}$	$\frac{6}{10}$	$\frac{3}{10}$	$1\frac{4}{5}$
$\frac{8}{10}$	$\frac{5}{10}$	$\frac{2}{10}$	$1\frac{1}{2}$
$\frac{7}{10}$	$\frac{4}{10}$	$\frac{1}{10}$	$1\frac{1}{5}$
$2\frac{2}{5}$	$1\frac{1}{2}$	$\frac{3}{5}$	

45.

$\frac{5}{10}$	$\frac{7}{10}$	$\frac{2}{10}$	$1\frac{2}{5}$
$\frac{3}{10}$	$\frac{9}{10}$	$\frac{4}{10}$	$1\frac{3}{5}$
$\frac{8}{10}$	$\frac{6}{10}$	$\frac{1}{10}$	$1\frac{1}{2}$
$1\frac{3}{5}$	$2\frac{1}{5}$	$\frac{7}{10}$	

46.

$\frac{3}{10}$	$\frac{8}{10}$	$\frac{7}{10}$	$1\frac{4}{5}$
$\frac{9}{10}$	$\frac{4}{10}$	$\frac{6}{10}$	$1\frac{9}{10}$
$\frac{1}{10}$	$\frac{2}{10}$	$\frac{5}{10}$	$\frac{4}{5}$
$1\frac{3}{10}$	$1\frac{2}{5}$	$1\frac{4}{5}$	

47.

$\frac{1}{10}$	$\frac{9}{10}$	$\frac{4}{10}$	$1\frac{2}{5}$
$\frac{5}{10}$	$\frac{2}{10}$	$\frac{6}{10}$	$1\frac{3}{10}$
$\frac{8}{10}$	$\frac{7}{10}$	$\frac{3}{10}$	$1\frac{4}{5}$
$1\frac{2}{5}$	$1\frac{4}{5}$	$1\frac{3}{10}$	

48.

$\frac{1}{10}$	$\frac{3}{10}$	$\frac{4}{10}$	$\frac{4}{5}$
$\frac{5}{10}$	$\frac{8}{10}$	$\frac{2}{10}$	$1\frac{1}{2}$
$\frac{7}{10}$	$\frac{6}{10}$	$\frac{9}{10}$	$2\frac{1}{5}$
$1\frac{3}{10}$	$1\frac{7}{10}$	$1\frac{1}{2}$	

49.

$\frac{9}{10}$	$\frac{2}{10}$	$\frac{5}{10}$	$1\frac{3}{5}$
$\frac{7}{10}$	$\frac{3}{10}$	$\frac{8}{10}$	$1\frac{4}{5}$
$\frac{1}{10}$	$\frac{4}{10}$	$\frac{6}{10}$	$1\frac{1}{10}$
$1\frac{7}{10}$	$\frac{9}{10}$	$1\frac{9}{10}$	

50.

$\frac{3}{10}$	$\frac{8}{10}$	$\frac{6}{10}$	$1\frac{7}{10}$
$\frac{5}{10}$	$\frac{9}{10}$	$\frac{1}{10}$	$1\frac{1}{2}$
$\frac{2}{10}$	$\frac{7}{10}$	$\frac{4}{10}$	$1\frac{3}{10}$
1	$2\frac{2}{5}$	$1\frac{1}{10}$	

51.

$\frac{2}{10}$	$\frac{6}{10}$	$\frac{1}{10}$	$\frac{9}{10}$
$\frac{9}{10}$	$\frac{5}{10}$	$\frac{8}{10}$	$2\frac{1}{5}$
$\frac{7}{10}$	$\frac{3}{10}$	$\frac{4}{10}$	$1\frac{2}{5}$
$1\frac{4}{5}$	$1\frac{2}{5}$	$1\frac{3}{10}$	

52.

$\frac{1}{10}$	$\frac{6}{10}$	$\frac{2}{10}$	$\frac{9}{10}$
$\frac{4}{10}$	$\frac{8}{10}$	$\frac{9}{10}$	$2\frac{1}{10}$
$\frac{5}{10}$	$\frac{3}{10}$	$\frac{7}{10}$	$1\frac{1}{2}$
1	$1\frac{7}{10}$	$1\frac{4}{5}$	

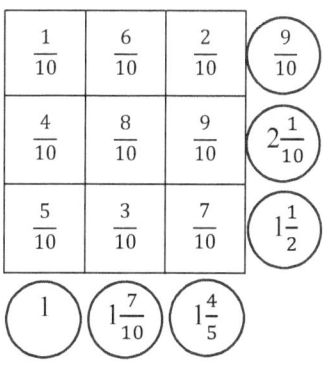

53.

$\frac{1}{10}$	$\frac{7}{10}$	$\frac{4}{10}$	$1\frac{1}{5}$
$\frac{2}{10}$	$\frac{8}{10}$	$\frac{6}{10}$	$1\frac{3}{5}$
$\frac{5}{10}$	$\frac{9}{10}$	$\frac{3}{10}$	$1\frac{7}{10}$
$\frac{4}{5}$	$2\frac{2}{5}$	$1\frac{3}{10}$	

54.

$\frac{8}{10}$	$\frac{5}{10}$	$\frac{6}{10}$	$1\frac{9}{10}$
$\frac{1}{10}$	$\frac{4}{10}$	$\frac{2}{10}$	$\frac{7}{10}$
$\frac{3}{10}$	$\frac{9}{10}$	$\frac{7}{10}$	$1\frac{9}{10}$
$1\frac{1}{5}$	$1\frac{4}{5}$	$1\frac{1}{2}$	

55.

$\frac{9}{10}$	$\frac{7}{10}$	$\frac{1}{10}$	$1\frac{7}{10}$
$\frac{4}{10}$	$\frac{2}{10}$	$\frac{6}{10}$	$1\frac{1}{5}$
$\frac{8}{10}$	$\frac{5}{10}$	$\frac{3}{10}$	$1\frac{3}{5}$
$2\frac{1}{10}$	$1\frac{2}{5}$	1	

56.

$\frac{4}{10}$	$\frac{1}{10}$	$\frac{9}{10}$	$1\frac{2}{5}$
$\frac{6}{10}$	$\frac{5}{10}$	$\frac{2}{10}$	$1\frac{3}{10}$
$\frac{3}{10}$	$\frac{8}{10}$	$\frac{7}{10}$	$1\frac{4}{5}$
$1\frac{3}{10}$	$1\frac{2}{5}$	$1\frac{4}{5}$	

57.

$\frac{6}{10}$	$\frac{5}{10}$	$\frac{9}{10}$	2
$\frac{1}{10}$	$\frac{4}{10}$	$\frac{7}{10}$	$1\frac{1}{5}$
$\frac{8}{10}$	$\frac{2}{10}$	$\frac{3}{10}$	$1\frac{3}{10}$
$1\frac{1}{2}$	$1\frac{1}{10}$	$1\frac{9}{10}$	

58.

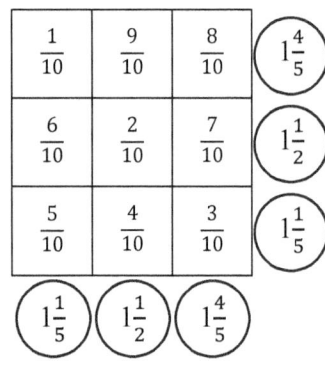

$\frac{1}{10}$	$\frac{9}{10}$	$\frac{8}{10}$	$1\frac{4}{5}$
$\frac{6}{10}$	$\frac{2}{10}$	$\frac{7}{10}$	$1\frac{1}{2}$
$\frac{5}{10}$	$\frac{4}{10}$	$\frac{3}{10}$	$1\frac{1}{5}$
$1\frac{1}{5}$	$1\frac{1}{2}$	$1\frac{4}{5}$	

59.

$\frac{6}{10}$	$\frac{8}{10}$	$\frac{9}{10}$	$2\frac{3}{10}$
$\frac{3}{10}$	$\frac{1}{10}$	$\frac{2}{10}$	$\frac{3}{5}$
$\frac{5}{10}$	$\frac{4}{10}$	$\frac{7}{10}$	$1\frac{3}{5}$
$1\frac{2}{5}$	$1\frac{3}{10}$	$1\frac{4}{5}$	

60.

$\frac{9}{10}$	$\frac{6}{10}$	$\frac{8}{10}$	$2\frac{3}{10}$
$\frac{3}{10}$	$\frac{2}{10}$	$\frac{4}{10}$	$\frac{9}{10}$
$\frac{7}{10}$	$\frac{5}{10}$	$\frac{1}{10}$	$1\frac{3}{10}$
$1\frac{9}{10}$	$1\frac{3}{10}$	$1\frac{3}{10}$	

61.

$\frac{9}{10}$	$\frac{3}{10}$	$\frac{6}{10}$	$1\frac{4}{5}$
$\frac{8}{10}$	$\frac{2}{10}$	$\frac{5}{10}$	$1\frac{1}{2}$
$\frac{7}{10}$	$\frac{1}{10}$	$\frac{4}{10}$	$1\frac{1}{5}$
$2\frac{2}{5}$	$\frac{3}{5}$	$1\frac{1}{2}$	

62.

$\frac{2}{10}$	$\frac{1}{10}$	$\frac{5}{10}$	$\frac{4}{5}$
$\frac{4}{10}$	$\frac{3}{10}$	$\frac{7}{10}$	$1\frac{2}{5}$
$\frac{6}{10}$	$\frac{9}{10}$	$\frac{8}{10}$	$2\frac{3}{10}$
$1\frac{1}{5}$	$1\frac{3}{10}$	2	

63.

$\frac{8}{10}$	$\frac{3}{10}$	$\frac{1}{10}$	$1\frac{1}{5}$
$\frac{4}{10}$	$\frac{6}{10}$	$\frac{9}{10}$	$1\frac{9}{10}$
$\frac{2}{10}$	$\frac{7}{10}$	$\frac{5}{10}$	$1\frac{2}{5}$
$1\frac{2}{5}$	$1\frac{3}{5}$	$1\frac{1}{2}$	

64.

$\frac{4}{10}$	$\frac{2}{10}$	$\frac{8}{10}$	$1\frac{2}{5}$
$\frac{3}{10}$	$\frac{6}{10}$	$\frac{1}{10}$	1
$\frac{9}{10}$	$\frac{7}{10}$	$\frac{5}{10}$	$2\frac{1}{10}$
$1\frac{3}{5}$	$1\frac{1}{2}$	$1\frac{2}{5}$	

65.

$\frac{3}{10}$	$\frac{4}{10}$	$\frac{5}{10}$	$1\frac{1}{5}$
$\frac{6}{10}$	$\frac{1}{10}$	$\frac{7}{10}$	$1\frac{2}{5}$
$\frac{8}{10}$	$\frac{9}{10}$	$\frac{2}{10}$	$1\frac{9}{10}$
$1\frac{7}{10}$	$1\frac{2}{5}$	$1\frac{2}{5}$	

66.

$\frac{7}{10}$	$\frac{4}{10}$	$\frac{1}{10}$	$1\frac{1}{5}$
$\frac{9}{10}$	$\frac{6}{10}$	$\frac{2}{10}$	$1\frac{7}{10}$
$\frac{5}{10}$	$\frac{3}{10}$	$\frac{8}{10}$	$1\frac{3}{5}$
$2\frac{1}{10}$	$1\frac{3}{10}$	$1\frac{1}{10}$	

67.

$\frac{6}{10}$	$\frac{2}{10}$	$\frac{7}{10}$	$1\frac{1}{2}$
$\frac{8}{10}$	$\frac{1}{10}$	$\frac{5}{10}$	$1\frac{2}{5}$
$\frac{4}{10}$	$\frac{9}{10}$	$\frac{3}{10}$	$1\frac{3}{5}$
$1\frac{4}{5}$	$1\frac{1}{5}$	$1\frac{1}{2}$	

68.

$\frac{7}{10}$	$\frac{6}{10}$	$\frac{9}{10}$	$2\frac{1}{5}$
$\frac{1}{10}$	$\frac{8}{10}$	$\frac{4}{10}$	$1\frac{3}{10}$
$\frac{2}{10}$	$\frac{5}{10}$	$\frac{3}{10}$	1
1	$1\frac{9}{10}$	$1\frac{3}{5}$	

69.

$\frac{9}{10}$	$\frac{8}{10}$	$\frac{6}{10}$	$2\frac{3}{10}$
$\frac{1}{10}$	$\frac{2}{10}$	$\frac{4}{10}$	$\frac{7}{10}$
$\frac{3}{10}$	$\frac{7}{10}$	$\frac{5}{10}$	$1\frac{1}{2}$
$1\frac{3}{10}$	$1\frac{7}{10}$	$1\frac{1}{2}$	

70.

$\frac{1}{10}$	$\frac{3}{10}$	$\frac{5}{10}$	$\frac{9}{10}$
$\frac{8}{10}$	$\frac{4}{10}$	$\frac{6}{10}$	$1\frac{4}{5}$
$\frac{2}{10}$	$\frac{7}{10}$	$\frac{9}{10}$	$1\frac{4}{5}$
$1\frac{1}{10}$	$1\frac{2}{5}$	2	

71.

$\frac{1}{10}$	$\frac{9}{10}$	$\frac{5}{10}$	$1\frac{1}{2}$
$\frac{4}{10}$	$\frac{3}{10}$	$\frac{8}{10}$	$1\frac{1}{2}$
$\frac{7}{10}$	$\frac{6}{10}$	$\frac{2}{10}$	$1\frac{1}{2}$
$1\frac{1}{5}$	$1\frac{4}{5}$	$1\frac{1}{2}$	

72.

$\frac{6}{10}$	$\frac{7}{10}$	$\frac{3}{10}$	$1\frac{3}{5}$
$\frac{5}{10}$	$\frac{9}{10}$	$\frac{2}{10}$	$1\frac{3}{5}$
$\frac{4}{10}$	$\frac{1}{10}$	$\frac{8}{10}$	$1\frac{3}{10}$
$1\frac{1}{2}$	$1\frac{7}{10}$	$1\frac{3}{10}$	

73.

$\frac{1}{10}$	$\frac{7}{10}$	$\frac{9}{10}$	$1\frac{7}{10}$
$\frac{5}{10}$	$\frac{2}{10}$	$\frac{8}{10}$	$1\frac{1}{2}$
$\frac{3}{10}$	$\frac{6}{10}$	$\frac{4}{10}$	$1\frac{3}{10}$
$\frac{9}{10}$	$1\frac{1}{2}$	$2\frac{1}{10}$	

74.

$\frac{5}{10}$	$\frac{4}{10}$	$\frac{2}{10}$	$1\frac{1}{10}$
$\frac{7}{10}$	$\frac{1}{10}$	$\frac{6}{10}$	$1\frac{2}{5}$
$\frac{9}{10}$	$\frac{8}{10}$	$\frac{3}{10}$	2
$2\frac{1}{10}$	$1\frac{3}{10}$	$1\frac{1}{10}$	

75.

$\frac{1}{10}$	$\frac{3}{10}$	$\frac{7}{10}$	$1\frac{1}{10}$
$\frac{5}{10}$	$\frac{2}{10}$	$\frac{4}{10}$	$1\frac{1}{10}$
$\frac{8}{10}$	$\frac{6}{10}$	$\frac{9}{10}$	$2\frac{3}{10}$
$1\frac{4}{10}$	$1\frac{1}{10}$	2	

76.

$\frac{7}{10}$	$\frac{5}{10}$	$\frac{3}{10}$	$1\frac{1}{2}$
$\frac{6}{10}$	$\frac{4}{10}$	$\frac{2}{10}$	$1\frac{1}{5}$
$\frac{9}{10}$	$\frac{8}{10}$	$\frac{1}{10}$	$1\frac{4}{5}$
$2\frac{1}{5}$	$1\frac{7}{10}$	$\frac{6}{10}$	

77.

$\frac{1}{10}$	$\frac{3}{10}$	$\frac{6}{10}$	1
$\frac{2}{10}$	$\frac{5}{10}$	$\frac{7}{10}$	$1\frac{2}{5}$
$\frac{4}{10}$	$\frac{8}{10}$	$\frac{9}{10}$	$2\frac{1}{10}$
$\frac{7}{10}$	$1\frac{3}{5}$	$2\frac{1}{5}$	

78.

$\frac{1}{10}$	$\frac{4}{10}$	$\frac{7}{10}$	$1\frac{1}{5}$
$\frac{5}{10}$	$\frac{2}{10}$	$\frac{3}{10}$	1
$\frac{9}{10}$	$\frac{6}{10}$	$\frac{8}{10}$	$2\frac{3}{10}$
$1\frac{1}{2}$	$1\frac{1}{5}$	$1\frac{4}{5}$	

The *MATHadazzles, Mind Stretch Puzzles* series include:

- Volume 1 Reasoning with Numbers

- Volume 2 Reasoning with Whole Numbers

- Volume 3 Reasoning with Integers

- Volume 4 Reasoning with Fractions

Contributors to Volumes 1, 2, and 3 are middle school teachers in the Greater Phoenix area.

Contributors to Volume 4 are middle school students.

www.ingramcontent.com/pod-product-compliance
Lightning Source LLC
Chambersburg PA
CBHW060401190526
45169CB00002B/704